HVAC
Equations, Data, and
Rules of Thumb

Arthur A. Bell Jr., PE

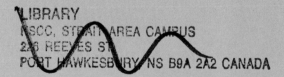
McGraw-Hill

New York   San Francisco   Washington, D.C.   Auckland   Bogotá
Caracas   Lisbon   London   Madrid   Mexico City   Milan
Montreal   New Delhi   San Juan   Singapore
Sydney   Tokyo   Toronto

Library of Congress Cataloging-in-Publication Data

Bell, Arthur A.
    HVAC : equations, data, and rules of thumb / Arthur A. Bell, Jr.
        p. cm.
    Includes bibliographical references and index.
    ISBN 0-07-136129-4
        1. Heating—Mathematics—Handbooks, manuals, etc.   2.
    Ventilation—Mathematics—Handbooks, manuals, etc.   3. Air
    conditioning—Mathematics—Handbooks, manuals, etc.   4.
    Mathematics—Formulae—Handbooks, manuals, etc.   I. Title.
    TH7225 .B45 2000
    697'.001'51—dc21   00-025029

McGraw-Hill

A Division of The McGraw·Hill Companies

5 6 7 8 9 0 DOC/DOC 0 6 5 4 3 2 1

ISBN 0-07-136129-4

The sponsoring editor for this book was Linda Ludewig, the editing supervisor was Frank Kotowski, Jr., and the production supervisor was Pamela A. Pelton.

It was set in Minion by North Market Street Graphics.

Printed and bound by R. R. Donnelley & Sons, Inc.

McGraw-Hill books are available at special quantity discounts to use as premiums and sales promotions, or for use in corporate training programs. For more information, please write to the Director of Special Sales, Professional Publishing, McGraw-Hill, Two Penn Plaza, New York, NY 10121-2298. Or contact your local bookstore.

    This book is printed on recycled, acid-free paper containing a minimum of 50% recycled, de-inked fiber.